No. 3025
$12.95

SCIENCE FOR YOU 112 ILLUSTRATED EXPERIMENTS

BOB BROWN

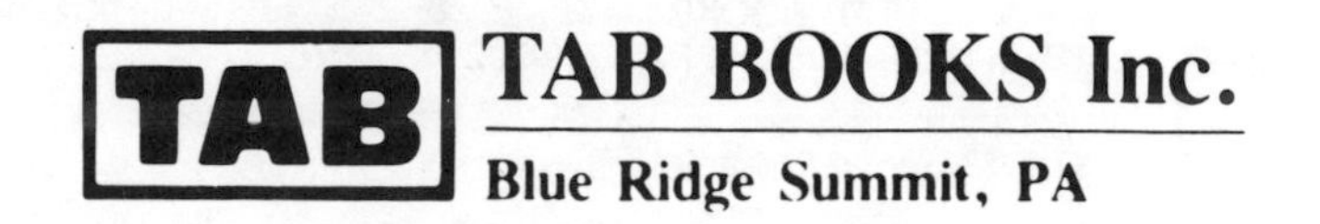

FIRST EDITION
FIRST PRINTING

Library of Congress Cataloging in Publication Data

Brown, Bob, 1907-
Science for you—112 illustrated experiments.

Includes index.
1. Science—Experiments—Juvenile literature.
I. Title. II. Title: Science for you—one hundred twelve illustrated experiments.
Q164.B843 1988 507'.8 88-8569
ISBN 0-8306-0325-5
ISBN 0-8306-9325-4 (pbk.)

Questions regarding the content of this book should be addressed to:

Reader Inquiry Branch
TAB BOOKS Inc.
Blue Ridge Summit, PA 17294-0214

Contents

Introduction

What is the happiest sound from the voice of a child? There are many, but one of the happiest is the exclamation of joy when the child claps hands and calls out for all to hear "it works, it works."

There are experiments for all—a huge reservoir to draw from. This collection holds happiness for young people. It takes the form of a science experiment book. We like to think this book meets the requirement. So, get out the sand, baking soda, sticks, wire, batteries, bottles, and boxes and have fun.

Chapter 1

Experiments with Electricity and Magnetism

PROBLEM:

Electrolytes.

NEEDED:

A battery, lamp and socket, tall glasses, wire, salt water, vinegar, household ammonia.

DO THIS:

Connect the battery and lamp as shown, so the bare ends of the wires extend down into the glass of water. No current flows to light the lamp. Dissolve salt in the water, and the salt water will conduct current so the lamp glows. Vinegar and household ammonia will make the water a conductor but not as effective as salt water.

HERE'S WHY:

Acids (vinegar), base (ammonia) and salts ionize the water (form charged particles) so the water conducts.

Pure water does not conduct because it is neither acid, base or salt. But most common tap water contains enough of some kind of soluble matter to make it at least a slight conductor. Oils are not conductors.

The illustration shows a toy train transformer instead of a battery.

PROBLEM:

Liberate chlorine.

NEEDED:

A D-cell battery, a large nail, wire and spring clips for connections, a jar, salt, a spoon, a toy train transformer.

DO THIS:

Cut the D cell and take the carbon out. It is the center rod. Clean it with soap and water. Clip a wire to the nail, one to the carbon, and both wires to the transformer. Immerse the nail and carbon in salt water.

WHAT HAPPENS:

Bubbles will rise from the carbon. Smell them! The odor will be something like kitchen bleach. It is the odor of chlorine, a gas liberated from the salt in the water (salt is sodium chloride). This small amount of chlorine is harmless. Large amounts would not be safe.

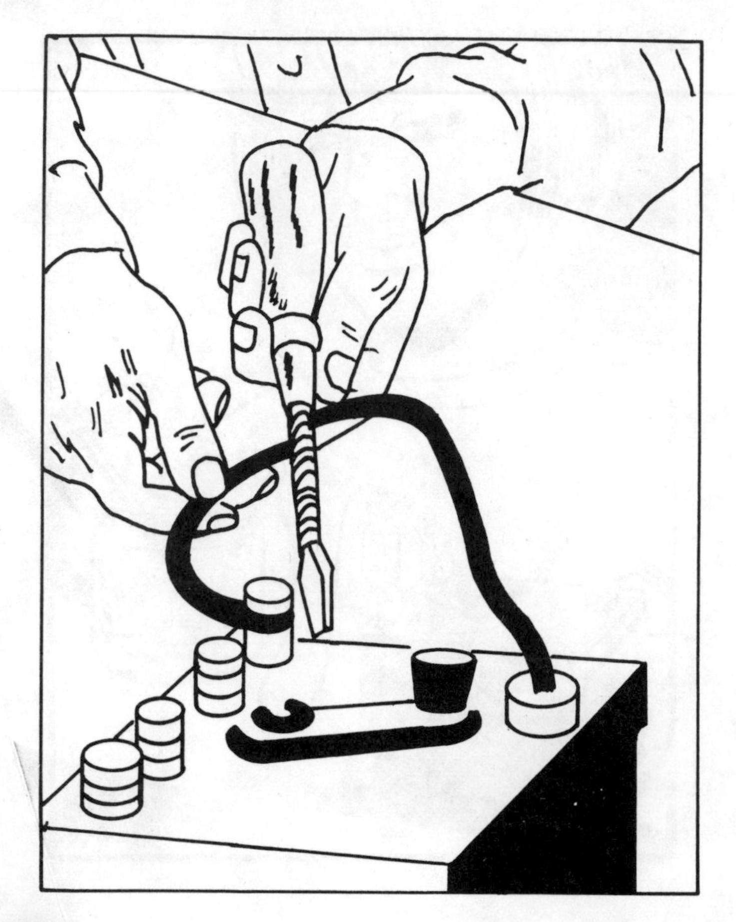

PROBLEM:

Magnetize a screwdriver.

NEEDED:

Insulated wire for connection to a large dry-cell battery.

DO THIS:

Wind several turns of wire around the screwdriver. Touch the bare ends of the wire to the battery instantly. There will be big sparks, and when the screwdriver is taken away it will be found to be magnetized.

HERE'S WHY:

Magnetism can be transferred from one magnet to the other. In this case, the turns of wire make up an electromagnet, and some of its magnetic strength is transferred into the steel of the screwdriver.

There is no shock to this experiment. But the wire will get hot if it is allowed to remain in contact with the battery.

The iron in some screwdrivers will remain magnetized a long time, but some screwdrivers cannot be magnetized well.

PROBLEM:

The darkening light bulb.

NEEDED:

A light bulb that has been in use a long time.

DO THIS:

Examine the bulb. The inside of the glass will have darkened at the upper part of the bulb.

HERE'S WHY:

The metal that makes up the hot filament tends to evaporate from the heat—slowly, of course. The evaporated metal particles circulate upward in the small amount of air or gas in the bulb and lodge on the inside of the glass.

PROBLEM:

Harmless electric sparks.

NEEDED:

A dry cell battery or toy train transformer, a file, some wire.

DO THIS:

Connect a wire to the file, and the other end of it to the battery. Connect the other wire to the other terminal of the battery. Then rub the free end of wire over the file, and sparks will fly!

COMMENT:

Such sparks will not burn the hand and will not set fire to ordinary substance. They could light gas, or gasoline, or other very combustible substances, but this is very unlikely.

The sparks are produced as current flows through the wire to the file then is stopped briefly as the contact is broken. Intense heat is produced in a very small area as the current is broken and arcs form. The arc is hot—what is seen is a shower of very tiny white hot particles of metal.

PROBLEM:

Magnetism (1).

NEEDED:

A strong magnet, some tacks, a glass jar.

DO THIS:

Place tacks in the jar, hold the magnet to the bottom of the jar, then invert it. The magnet holds the tacks to the top of the jar. Move the magnet away and the tacks drop.

COMMENT:

Magnetism passes through the glass to attract the tacks and hold them in a mass at the top of the jar. As the magnet is removed, the magnetic strength diminishes and lets the tacks fall.

PROBLEM:

Magnetism (2).

NEEDED:

An old record player or a toy record player, thin cardboard, a small magnet or two, sticky tape, some tacks or small nails.

DO THIS:

Tape the magnet to the turntable. Place a few tracks on the card, start the turntable, and lower the card and tacks almost to it.

WHAT HAPPENS:

The magnet, exerting its pull through the card, will pull the tacks round and round. Try small nails instead of tacks. (Don't play with the family's expensive stereo.)

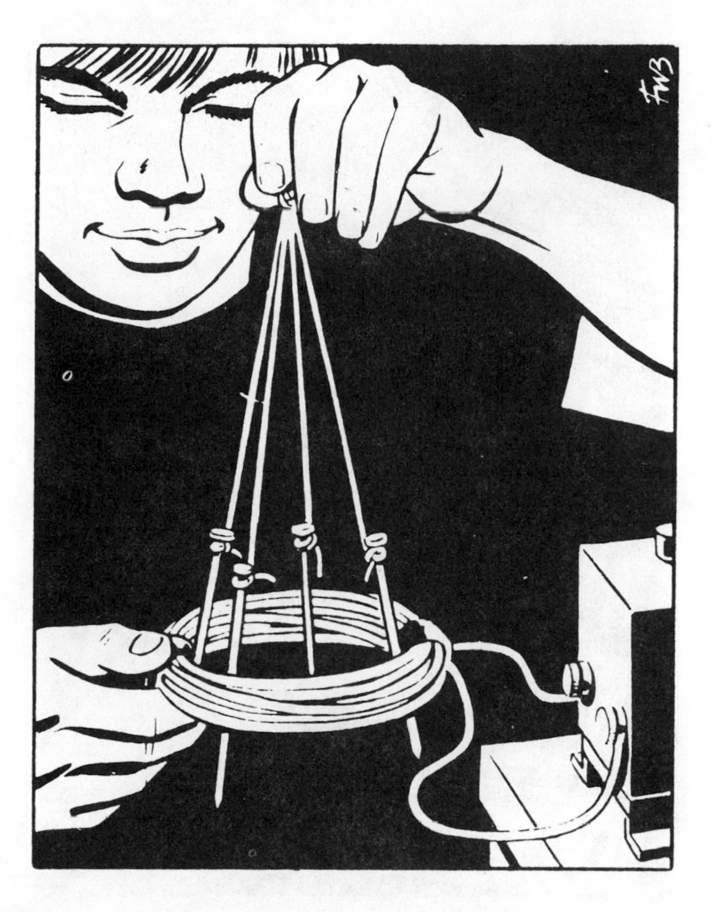

PROBLEM:

An electric trick.

NEEDED:

A toy train's electrical transformer, a coil of insulated wire, nails and string. The transformer should give ac (alternating current).

DO THIS:

Tie each nail to a string, then tie the other ends of the strings together. Suspend the nails in the coil then turn on the transformer. The nails will repel each other and move quickly toward the coil.

HERE'S WHY:

When the transformer is turned on, a magnetic field is formed by the coil. Through the transformer's effect, a current is induced in each nail, and they become magnetized. The nails repel each other according to the rule "like poles repel and unlike poles attract."

PROBLEM:

Electric "glue."

NEEDED:

Smooth sheet of newspaper, a table top.

DO THIS:

Place the paper on the table and rub it briskly with the hand. It should stick to the table as if glued down. If a corner of the paper is raised, it will flop down again. Try rubbing the paper with a pencil and various kinds of cloth.

WHAT HAPPENS:

Rubbing causes the paper to pick up an electrostatic charge. The table top is charged oppositely and attracts the paper, because opposite charges attract each other. The paper is not a good conductor of charges and may keep a charge for several minutes before being discharged by ions in the air.

As in most static experiments, this may not work unless the humidity in the air is low.

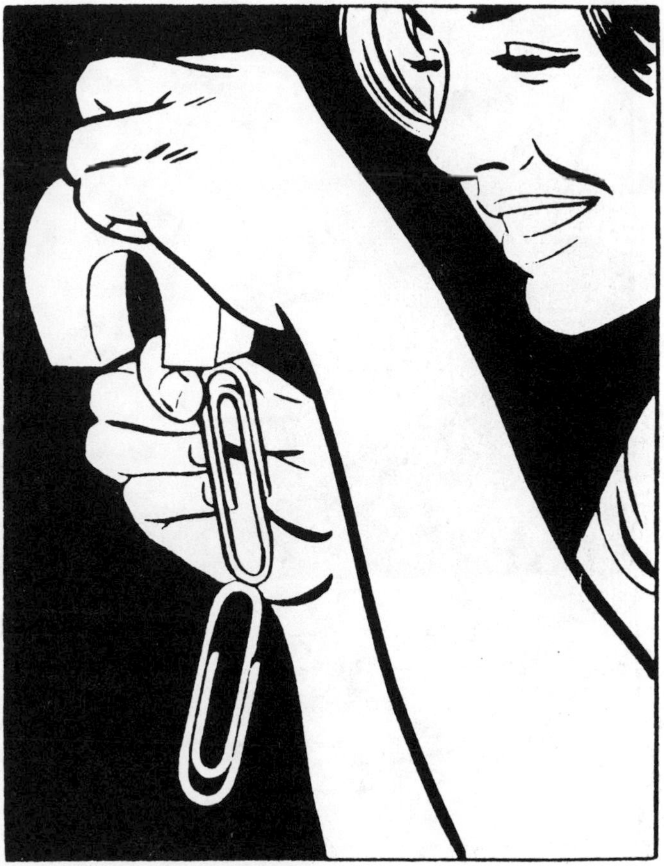

PROBLEM:

Magnetism by induction.

NEEDED:

A magnet, two paper clips.

DO THIS:

Suspend both clips from the magnet as shown. Pull the upper one away from the magnet, and the bottom one falls.

HERE'S WHY:

The first clip becomes a magnet as it touches the permanent magnet. It then can hold the second clip up. But when the first clip is pulled from the magnet it no longer is a magnet itself, and the lower clip falls.

Paper clips are soft iron. They do not hold magnetism permanently as steel does.

PROBLEM:

Does the compass needle point north?

COMMENT:

A compass points in the magnetic north-south direction, which may be affected by electric wires and pieces of iron nearby. The true, nonmagnetic north-south direction depends on the axis of rotation of the earth and is rarely the same as magnetic north-south.

In the northeastern states a magnetic compass will point about 20 degrees west of true north, while in the northwestern states it will point about 20 degrees east of true north.

The cause of the earth's magnetic field is not known.

PROBLEM:

Sparks.

NEEDED:

Long-play record, wool, kitchen foil, a table top.

DO THIS:

Cut a five-inch disk from the foil. Place the record on the table, rub it briskly with the wool, drop the foil on it, lift the record and foil from the table, and bring a finger to the foil. A spark should be seen and heard between the finger and the foil.

HERE'S WHY:

Rubbing the record produces a static electric charge on it.

This experiment must be performed in a room where humidity is low. Winter is best for this experiment.

PROBLEM:

Snap, crackle, pop.

NEEDED:

A comb, puffed rice, low humidity.

DO THIS:

Put the puffed rice into a bowl. Rub the comb briskly with a piece of silk to add a charge of static electricity. Stick the comb down into the cereal, take it out, and watch.

WHAT SHOULD HAPPEN:

The grains should stick to the comb momentarily; then, as they take on the same charge as the comb, they are repelled quickly. The grains give up electrons to the comb until the charges on comb and grains are the same.

This experiment must be performed when the humidity is low. Winter is best.

Chapter 2

Experiments with Liquids

PROBLEM:

Floating.

NEEDED:

A jar lid, a needle, a vessel of water.

DO THIS:

Float the lid on the water surface. Lower a needle carefully and it also will float on the water, but for a different reason.

HERE'S WHY:

The lid contains air, which makes it light enough to float. The surface tension of the water holds the needle up.

Push the lid down until water gets into it, and it will be heavy enough to sink. Push the end of the needle down, breaking the surface tension, and it, too, will sink.

PROBLEM:

Surface tension.

NEEDED:

A flat white plate, water, food color or ink, rubbing alcohol.

DO THIS:

Put a very thin layer of colored water on the plate. Get a drop of alcohol on the fingertip and touch it to the middle of the plate. The colored water will rush away from the finger.

HERE'S WHY:

Surface tension, like a thin sheet, covers the water. The surface tension of alcohol is less, so as the alcohol touches the water the greater surface tension of the water pulls the alcohol away from the finger.

PROBLEM:

Surface tension.

NEEDED:

Sewing thread (cotton is best), water, detergent.

DO THIS:

Make a loop of sewing thread. Wet it, then float it on the surface of the water.

WHAT HAPPENS:

The thread will form an irregular circle as it floats. Put a drop of detergent inside the loop, and it springs into a perfect circle shape.

HERE'S WHY:

Surface tension on the water is like a flexible sheet stretched across it. The pull is equal on all the water surface. But the detergent breaks or weakens the surface tension on the water inside the loop, and the greater tension outside the loop pulls the thread into the circular shape.

PROBLEM:

Oil in the water.

NEEDED:

A medicine dropper, a glass of water, rubbing alcohol, oil.

DO THIS:

Put some alcohol into the dropper and release it slowly under the surface of the water in the glass. Nothing happens.

Release drops of oil the same way. The oil forms round drops, because of surface tension, and when it reaches the surface of the water it still remains in drops, although gravity will flatten them out somewhat.

The alcohol dissolves in the water; the oil does not.

PROBLEM:

The vortex.

NEEDED:

Water in a container and a spoon.

DO THIS:

Stir the water and it will climb up the walls of the container, leaving a "hole" in the middle. If a blender is used instead of a spoon, the water may be made to swirl so fast it flies out of the top of the container.

HERE'S WHY:

The water tends to fly away from the center by centrifugal force. The hole at the bottom is not as large as that at the top, and this is because the pressure of the water increases with depth. The weight of the water above prevents the lower water from moving outward so much.

PROBLEM:

Water "glue."

NEEDED:

A spigot, a large spoon.

DO THIS:

Hold the rounded part of the spoon in a smooth flow of water. Note that the water stream seems to hold the spoon as if glued to it.

HERE'S WHY:

This is an example of the Coanda effect. The water next to the spoon develops a slight motion that reduces the pressure, and atmospheric pressure on the side of the flow opposite the spoon tends to hold the stream against the spoon. The "teapot effect," in which tea tends to run down the side of the cup, is seen here.

PROBLEM:

Flotation.

NEEDED:

Two glasses that fit into each other, a small amount of water.

DO THIS:

See how little water it takes in the lower glass to make the upper glass float.

EXPLANATION:

The shape of the container determines how little water it takes to float a heavy object. If the item to be floated fits closely inside the container, there is very little space—in this case between the two glasses—for the displacement of water. Hence, the interior glass will float on very little water. Look up the Archimedes principle.

PROBLEM:

Buoyancy.

NEEDED:

Two rubber balloons, two containers of water, warm and cold.

DO THIS:

Fill two balloons with cold water, tie them and place one in the container of cold water. Now place the other in warm water. You will see that the first balloon sinks.

WHAT HAPPENS:

Because cold water is heavier than warm water, the balloon filled with cold water will always sink when placed in warm water. You will see if the conditions are reversed that a balloon filled with warm water will float in a container filled with cold water.

PROBLEM:

Floating.

NEEDED:

Aluminum foil and a pan of water.

DO THIS:

Wad the foil into a ball and drop it into the water. Form a piece of foil into a boat and drop it, too, into the water.

COMMENT:

Wad the foil into a ball, and there are so many air spaces in the ball that it floats. A boat made of foil naturally floats as all well-made boats do. A sheet of foil, if pushed under the water, will sink.

PROBLEM:

Water purity.

NEEDED:

Tap water, distilled water, two drinking glasses.

DO THIS:

Put equal amounts of tap water and distilled water in the glasses, and set them aside for the water to evaporate.

WHAT HAPPENS:

Dirty rings form on the glass that held the tap water. The rings indicate that the water contained impurities. A chemist can learn what the various impurities are. If the glass containing distilled water shows dirty rings, we may conclude the impurities there dropped into the water from the air. Impurities in tap water are usually harmless to the body.

Chapter 3

Experiments with Gases

PROBLEM:

Air pressure.

NEEDED:

A plastic bottle, a square of card or thick paper, a vessel of water.

DO THIS:

Place the paper over the mouth of the bottle and insert the bottle into the water. Go easy, and the paper will remain on the bottle mouth even when the bottle is turned upside down. If it tends to slip off, squeeze a little air out of the bottle, and the paper should stick tightly. Squeeze again, and the paper will slide off.

HERE'S WHY:

The pressure on the water side of the paper is greater than the pressure on the air side of the paper. The force due to this slightly greater pressure holds the paper tightly against the mouth of the bottle.

PROBLEM:

Steam.

NEEDED:

Observation.

NOTE:

On a cold day steam comes from the exhaust pipe of the automobile.

COMMENT:

The exhaust contains warm, moist particles, much of it water that is released as steam from the burning fuel. When the exhaust mixture meets cold air much of it condenses, forming a cloud or fog.

Hot steam is invisible, but cold air condenses it into small visible droplets that reflect light in such a way that they appear white.

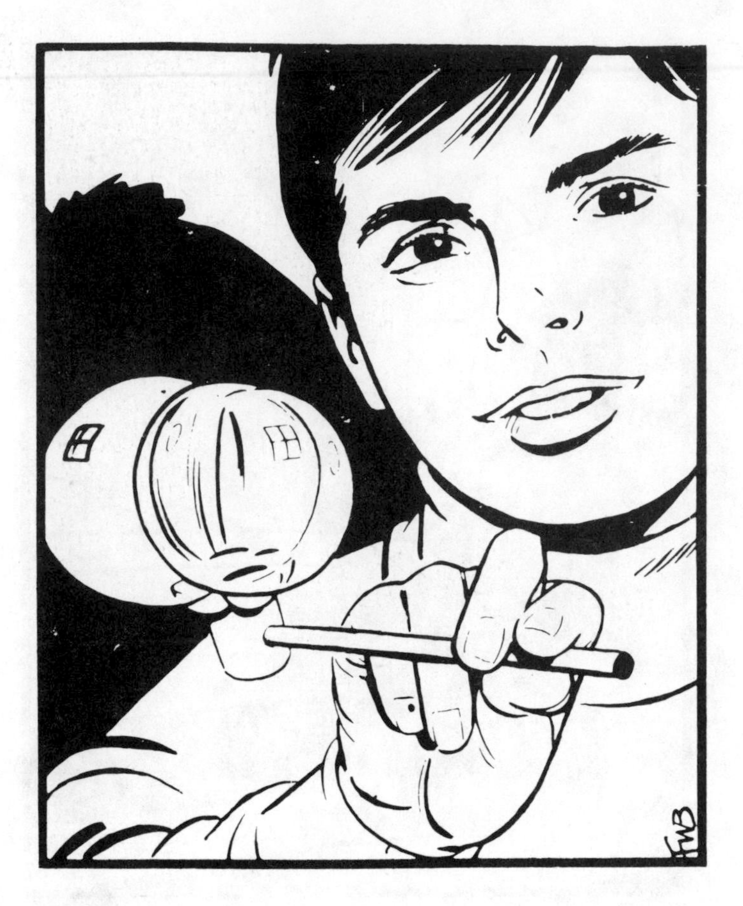

PROBLEM:

Bubble shapes.

NEEDED:

Bubble blowing equipment.

DO THIS:

Blow a regular bubble; it is almost round, distorted somewhat by the breeze and the pull of gravity. Blow into the solution so several bubbles form together. The roundness does not apply where bubbles join.

HERE'S WHY:

The single bubble is round because of the almost equal effect of surface tension all around the surfaces of the bubble. This surface tension makes the air pressure inside the bubble slightly greater than that outside.

When bubbles are joined the surface between them may be flat or may extend into one of the bubbles. It all depends on the pull of surface tension, which continually pulls toward the smallest area.

PROBLEM:

Bubble images.

NEEDED:

Bubble solution.

DO THIS:

Hold a bubble still in front of the face. Notice that there is a reflection from the front part of the bubble, right side up, and another reflection from the back part of the bubble, upside down.

HERE'S WHY:

The surfaces of water making up the bubble make two mirrors. The front of the bubble is a convex reflector, which leaves the image right side up. The back of the bubble makes a concave mirror, which inverts the image.

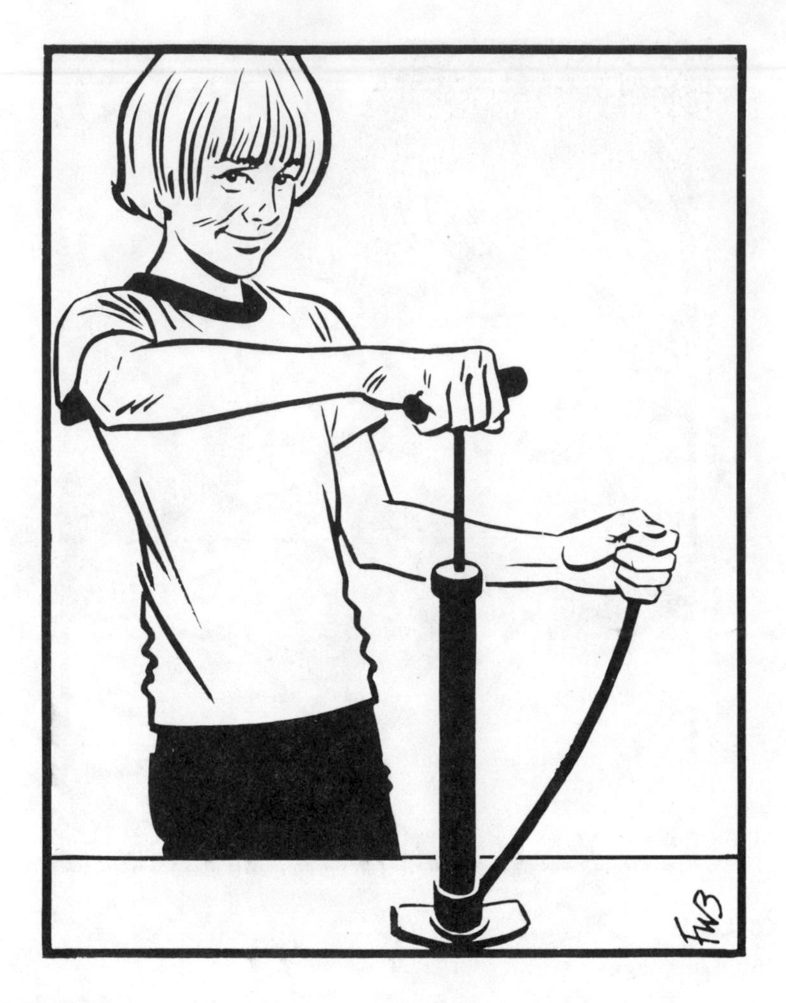

PROBLEM:

Elastic gases.

NEEDED:

A hand tire pump.

DO THIS:

Hold a finger over the pump hose, and let someone push the plunger down. The plunger will return quickly to its original height when the downward push is discontinued.

HERE'S WHY:

Gases are elastic. That means that they return to their original volume after being compressed if no gas is lost in the operation.

One person or two may take part in this experiment.

PROBLEM:

Friction and air.

NEEDED:

A long rubber balloon and a wide mouth vinegar jug.

DO THIS:

Challenge a friend to lift the jug with the balloon. Then show him how: put the balloon into the jug, blow it up, and lift. It is easy.

HERE'S WHY:

When the rubber expands against the sides of the jug the friction created is probably sufficient to lift the jug, even though the jug may be full of water. Then, too, as the balloon is pulled outward the air pressure under it is reduced, so the greater pressure of the atmosphere, pushing upward on the bottom of the glass, helps prevent the balloon from being pulled out.

PROBLEM:

Blow a square bubble.

NEEDED:

A wire and some dishwashing detergent.

DO THIS:

Bend the wire into a square with a handle. Dip it into the detergent and lift it out so a detergent film fills the opening. Blow the film outward. The bubble comes out not square-shaped, but spherical.

HERE'S WHY:

Because of surface tension (the attraction of the molecules of the bubble for each other), the bubbles will be pulled into a sphere rather than a square-shaped object. A sphere has the smallest surface area of any shape, so the molecules automatically form that shape.

Chapter 4

Experiments with Sound and Other Vibrations

PROBLEM:

Sensitive ears.

NEEDED:

A rubber tube, a friend.

DO THIS:

Mark the tube at the half-length point. Put the ends into your ears. Have the friend scratch the tube with a fingernail or a comb, and you can tell which side of the middle is scratched.

COMMENT:

The scratch sound flows both directions from the point scratched, but the ears are very sensitive, and can tell which side is scratched, even though the point is near the center of the tube. Of course, the tube should be hanging behind the back when the scratching is done, so vision is not involved.

PROBLEM:

Sound for fishes.

NEEDED:

Go swimming.

DO THIS:

Have a friend take two stones into the water some distance from you. Have him bang them together in the air, and see how loud the sound is. Then have him place the stones underwater, and bang them together while your head is underwater. The sound will be much louder.

HERE'S WHY:

Sound exists in the form of waves, and the waves travel better in water than in air. (Solids carry sound better than air as a rule. American Indians are said to have listened for distant hoofbeats by placing an ear to the ground.)

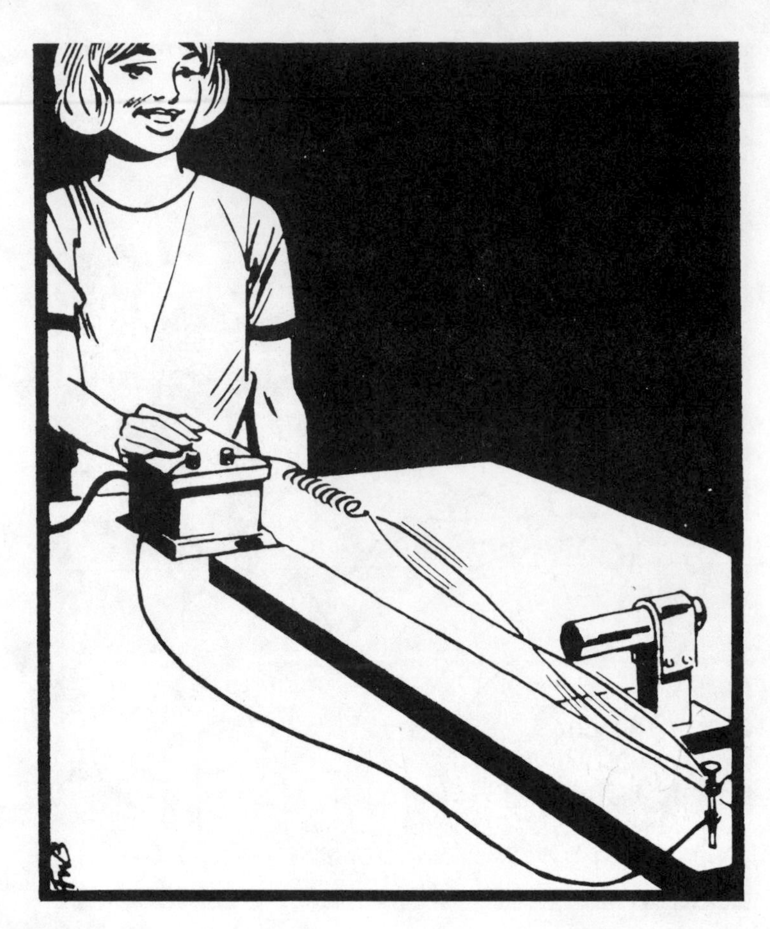

PROBLEM:

The singing wire.

NEEDED:

Copper wire, a strong magnet, a toy transformer and supports. (A spring is good but not necessary.)

DO THIS:

Stretch a few feet of the wire between supports (nails will do) and hold it with a spring or tighten it by hand around a nail. Connect the wire to the transformer, and move the magnet along the wire until the wire vibrates. The position of the magnet and tension on the wire will control the vibration.

WHAT HAPPENS:

The wire carries an alternating current, producing magnetic lines of force that interact with the lines from the permanent magnet, producing alternately attraction and repulsion of the wire.

The wire vibrates in accordance with the rules regarding vibration of a string, showing one or more nodes. The vibration may produce an audible musical sound.

PROBLEM:

Binaural sounds.

NEEDED:

A friend, two spoons (one suspended on a string).

DO THIS:

Have the friend sit in the middle of the room and hold one hand tightly against an ear. Hold spoon by the string, and tap it lightly with the other. Have the friend guess which part of the room the sound comes from. (Move around silently while making the sounds.)

WHAT HAPPENS:

We are used to hearing with two ears, "binaural" hearing. Sounds do not reach both ears at the same time, and this gives us information as to the direction of the soundmaker. This sense is lost if hearing is by one ear only.

We may "cock" our heads to better determine the direction of the sound.

PROBLEM:

How do we whistle?

NEEDED:

Puckered lips.

DO THIS:

Whistle through the lips, in the common way. Try to explain what makes the whistle.

COMMENT:

The sound is a "hole tone" caused by air meeting a hole made by the lips. There is a resonating cavity, the mouth, that adds to the sound.

There is some mystery to the common whistle. Dr. Jearl Walker, of Cleveland State University, possibly the most discussed physics writer of the day, says, "The details of the air flow do not seem to be worked out." The quote is from his book "The Flying Circus of Physics" published by John Wiley & Sons, Inc.

PROBLEM:

Door chimes.

NEEDED:

A wire coat hanger on a door.

DO THIS:

Close the door. The wire banging on the wood will produce "chimes."

COMMENT:

As the wire strikes the door it is set into vibration. Without the door the vibrations are low in intensity. The door, however, acts as a sounding board, amplifying and changing the effect, making the sound more mellow.

(Five collections of these scientific tricks are available in book form. Get a list from Bob Brown, 20 Vandalia, Asheville, NC 28806. Send a stamped, self-addressed envelope.)

PROBLEM:

Bottle noise.

NEEDED:

A bottle and a faucet.

DO THIS:

Run water into the bottle; the pitch of the bubbly noise gets higher. Pour water out, and the pitch of the noise gets lower.

HERE'S WHY:

Many frequencies of sound are made by the bubbling water, but those that excite the air column in the bottle are heard more clearly, as the air column resonates. The resonant frequencies are determined by the volume of air in the column, or length of the air column, and as the air volume increases the frequency increases.

PROBLEM:

Fry-pan chimes.

NEEDED:

A fry pan, tableware, hard cord, sticky tape.

DO THIS:

Tie each utensil to a string, loop the other ends of the strings into the pan, and hold them with tape. When the utensils jangle, the sound is pleasant as it comes from the pan.

HERE'S WHY:

Only the tableware will jangle if the strings are held in the hand. But if the strings hang from the pan, the vibrations of the tableware go up the strings, making the pan vibrate also. The sound coming from the pan is intensified and mellowed. The pan acts as a sounding board.

PROBLEM:

Rip and tear.

NEEDED:

A piece of cloth.

DO THIS:

Rip or tear the cloth. The faster it is torn, the higher the pitch of the sound made.

HERE'S WHY:

Each time a thread breaks it sets up a vibration of the air around it. If the threads are broken at a faster rate, the vibrations come faster and the pitch of the sound is higher.

PROBLEM:

Vibrations.

NEEDED:

A sheet of paper.

DO THIS:

Hold the paper in front of the face, hum with a loud voice, and the paper will be felt to vibrate although no air will be blowing through it.

HERE'S WHY:

The paper shakes because sound waves, and not a breeze, are blowing on it. Sound energy is being transferred to the paper to cause the vibration.

PROBLEM:

Try it!

NEEDED:

A round balloon, an alarm clock or watch.

DO THIS:

Blow up the balloon, and hold it between the clock and the ear. Notice how distinct the ticking is. Remove the balloon and see if the ticking is less loud.

COMMENT:

Carbon dioxide from the breath is heavier than air, and the balloon containing it is supposed to act as a converging lens to the sound waves, focusing them to the ear. The small amount of carbon dioxide in the balloon does not make any appreciable difference in the sound intensity.

To get more carbon dioxide into the balloon, mix baking soda and vinegar in a bottle and let the gas go into the balloon. The gas may be concentrated enough to produce the converging lens effect.

PROBLEM:

Funnel "music"

NEEDED:

A funnel (try a glass and a metal funnel).

DO THIS:

Put the funnel to the lips, and blow into the funnel by making the lips vibrate. You have a "musical" instrument.

HERE'S WHY:

The vibration of the lips sets the column of air in the funnel vibrating, and this is what is heard. The size and shape of the funnel as well as the lip movement itself determine the sound that comes out.

Several real musical instruments operate on this principle, including the trumpet and tuba. Some have valves to vary the length of the tubing through which the sound waves come.

PROBLEM:

The silence of snow.

NEEDED:

The out-of-doors after a snowfall.

DO THIS:

Listen to speech and other sounds. They are soft, weak and unusual. Snow is water. Rain is water. Why doesn't a rainfall produce the same result?

HERE'S WHY:

The snow is in crystals that have millions of spaces that absorb the sounds. The spaces are between flakes and inside the flakes themselves. Rain or other water does not have such spaces, and therefore reflects sound much better than snow.

PROBLEM:

Directed echo.

NEEDED:

Two cardboard tubes, a solid wall, an alarm clock.

DO THIS:

Place the tubes against the wall, so they make a 45-degree angle. Hold the clock at the end of one tube. Hold an ear at the end of the other. Listen to the ticking of the clock through the tube and outside the tube.

WHAT HAPPENS:

Ordinarily, the sound of the ticking spreads in all directions, becoming weaker as it moves from the source. The sound vibrations entering the tube tend to pass through the tube rather than scatter. If the tubes are held at the correct angle at the wall, some sound is reflected from one tube through the other.

PROBLEM:

Sound insulation.

NEEDED:

A styrofoam ice container, washcloths, an alarm clock.

DO THIS:

Make the clock alarm, place it into the container, put the lid on, and see how much less noisy the clock is. Take the clock out, place four washcloths in the bottom of the container, replace the clock, and the sound seems much less noisy.

HERE'S WHY:

When the clock is placed on the bottom of the container, the vibrations of the clock are transferred to the styrofoam and it acts as a sounding board to give off sound of its own. The washcloths help deaden that sound, and the walls of the container serve as sound insulators to diminish the total sound.

PROBLEM:

Resonant frequency.

NEEDED:

An oatmeal box, a sharp knife, sand.

DO THIS:

Cut a hole in the side of the box 2 inches from the end. Put a few pinches of sand on the end of the box. Hold the box to the mouth and hum into the hole. Begin with high notes and go down the scale. At one point, the grains of sand will dance up and down violently.

HERE'S WHY:

The end of the box has a natural vibration rate called the resonant frequency. When the voice sound is that frequency, it sets the box lid vibrating. Other frequencies do not make the lid vibrate appreciably.

PROBLEM:

Little echoes.

NEEDED:

A glass, a jar, a metal pan or bucket.

DO THIS:

Hum or speak into the mouth of each vessel, or a few inches from it. Echoes can be heard.

COMMENT:

Sound waves are reflected from hard surfaces, and in the case of the hard-surface vessels, the waves are reflected back and forth many times. The echoes here are quite different from those heard from a distant rock cliff, but they are echoes just the same.

PROBLEM:

Sound conduction.

NEEDED:

A watch, a table, a funnel.

DO THIS:

Place the watch on the table, and its sound may be heard faintly. Hold the funnel to the ear and see whether the sound is louder. Now place the ear against the table top and see whether the sound is louder coming through the wood.

COMMENT:

The sound can come through the air to the ear, but is likely to be louder when the funnel is used because the funnel concentrates some of the sound waves into the ear. The vibrations can travel through the wood of the table top to the ear.

PROBLEM:

Closed pipe.

NEEDED:

A vacuum cleaner tube or golf tube. A cardboard tube may work.

DO THIS:

Hold the tube *against* the ear and listen for any sound. Then move it *away* from the ear a fraction of an inch; the same sounds have a higher pitch.

HERE'S WHY:

This is the pipe organ principle. Resonant sound vibrations in a tube closed by the ear have a lower pitch than the vibrations in a tube that has both ends open, even though it is the same length.

And where do the sounds come from in this experiment? From the environment, just like the sounds that seem to come from a sea shell. The experiment will not work in a very quiet room.

PROBLEM:

Big bang.

NEEDED:

A paper bag.

DO THIS:

Fill the bag with air, crush it quickly between the hands, and listen for the "bang."

WHAT HAPPENS:

As the hands come together, the air in the bag is compressed quickly. The sound is heard as the compressed air is released and assumes regular atmospheric pressure.

PROBLEM:

The bull roarer.

NEEDED:

A foot-long strip of thin wood with a hole drilled in one end, a string.

DO THIS:

Tie a 2-foot string into the hole end of the wood. Hold the end of the string and whirl the stick around in front or over the head.

WHAT HAPPENS:

As the stick moves through the air, it catches in the air and begins whirling. As it whirls, it catches air in irregular gulps, causing the roaring waves of sound.

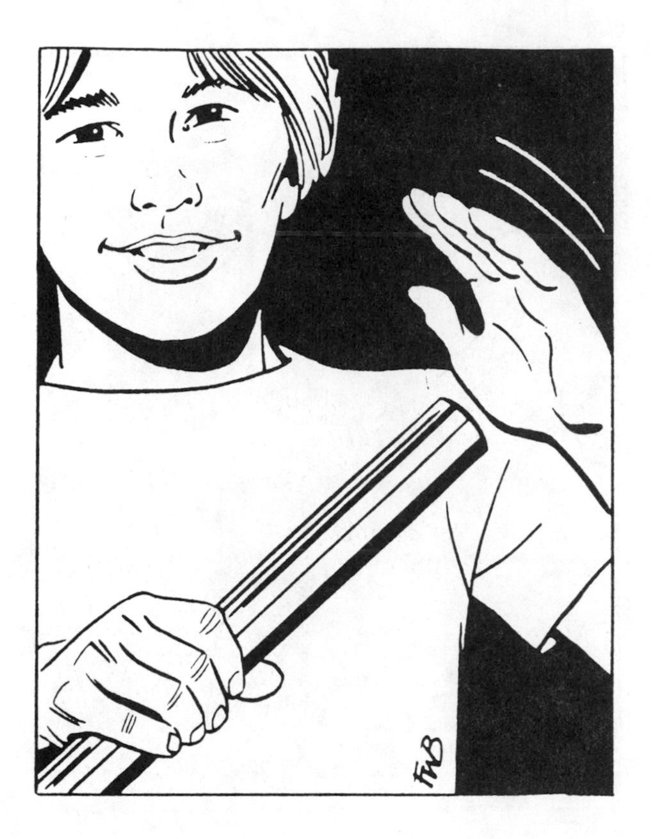

PROBLEM:

Sound in a tube.

NEEDED:

A two-section vacuum cleaner tube.

DO THIS:

Pat the end of one section of tube with the palm of the hand. You will hear a sound. Put the two sections together and pat; the sound is of a different pitch.

HERE'S WHY:

The air column in the tube will vibrate in resonance with certain sound frequencies, depending on the length of the tube, whether both ends are open, and other factors. (In resonance, the vibrations move together, each contributing strength to the other.)

PROBLEM:

Flower pot chimes.

NEEDED:

Flower pots and short ropes.

DO THIS:

Hang the pots on the ropes so they don't touch each other. Tap them with a spoon, and chimes of a deep muffled tone will be heard.

HERE'S WHY:

The pots will vibrate when struck and will make a muffled sound because they are thick and not very dense. Their porous and thick walls tend to deaden or absorb the higher pitches.

PROBLEM:

Music Lesson.

NEEDED:

A wire stretched tightly.

DO THIS:

Pluck the wire and hear the note produced. Hold your finger on the middle of the wire, pluck it again, and the tone will be an octave higher, because each half of the string vibrates faster than the entire length vibrated before. Hold another finger a third of the way across the wire, and the wire will vibrate in three sections, producing a note an octave and a fifth higher than the first note that you produced.

COMMENT:

The number and strength of overtones give an instrument its tone quality.

PROBLEM:

Resonance.

NEEDED:

Two vacuum cleaner tubes or other tubes.

DO THIS:

Clean the tubes. Hold one to your mouth and hum into it, varying the pitch. At one point a condition called the resonant frequency will be reached, and the tube will be felt to vibrate.

HERE'S WHY:

The resonant frequency of a column of air in a tube depends on the speed of sound in air and the length of the tube. A longer tube has a lower-pitched resonant frequency; when you hum into two tubes placed together end to end, you will have to hum at a lower pitch to make the tubes vibrate.

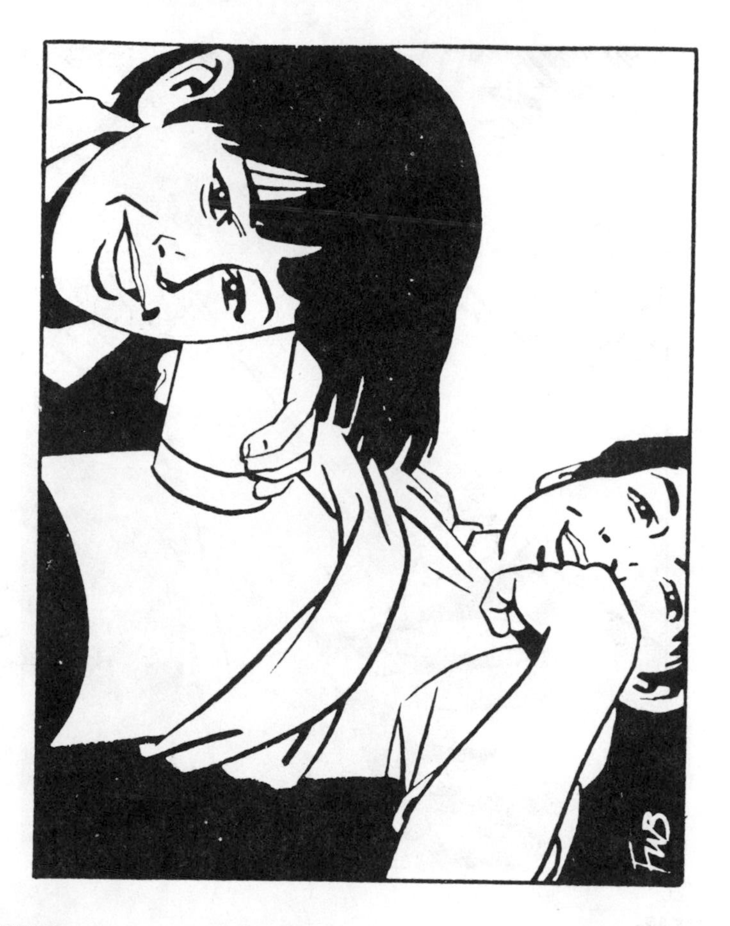

PROBLEM:

Hear heartbeats.

NEEDED:

A paper cup and a knife.

DO THIS:

Cut the bottom out of the cup. Hold the hole in the bottom to your ear and hold the other end to someone's chest and listen for the sound of his or her heartbeat.

WHAT HAPPENS:

The cup concentrates some of the sound waves, making them louder as they reach the ear. Actually no cup or other device is necessary to hear heartbeats if the ear is held against the chest and the room is quiet.

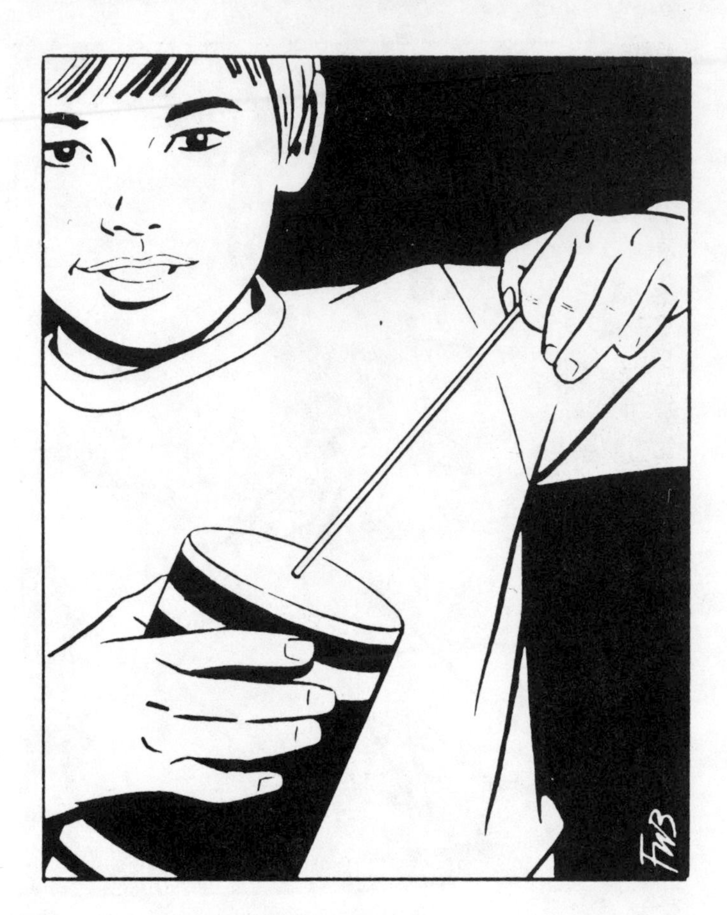

PROBLEM:

The growling dog.

NEEDED:

An oatmeal box, a cotton cord.

DO THIS:

Make a hole in the bottom of the box and push the cord through it. Make a knot in the cord to hold it in the hole. Hold the box in one hand and pull the other hand down the string, allowing a fingernail to rub against the cord. Leave the other end of the box open.

WHAT HAPPENS:

As the fingernail is pulled along the cord, it makes vibrations in the cord. The vibrations are transmitted to the box and make the air in the box vibrate. The vibrations create a growl. This growling sound can be rather realistic.

PROBLEM:

Feel the voice.

NEEDED:

Observation.

DO THIS:

Feel various parts of the throat while humming. You can feel strong vibrations when you touch the area of the vocal cords.

COMMENT:

The voice is produced as air passes through the vocal cords in a particular way. Animal sounds differ from one another because of the difference in the structure of their vocal cords.

Chapter 5

Experiments with Seeds and Plants

PROBLEM:

Grow sprouts.

NEEDED:

A quart jar, nylon net or cheesecloth, rubber band, seeds or beans that have not been treated with chemicals, water.

DO THIS:

Put one-fourth cup of seeds into the jar, pour one cup of water over them. Cover the jar with the cloth or net, holding it on with a rubber band. Place the jar in a dark place overnight. Pour off the soaking water, rinse the seeds and drain. Tilt the jar top-side down so excess water will drain. Rinse twice a day for three days or until sprouts are ½ to 1 inch long. Refrigerate in a closed container to increase the vitamin C. The sprouts are delicious in salads or soups.

The first soaking water should be saved for soups. Garden shop or seed store seeds should not be used; they may be treated with chemicals. Get lentils, soybeans, mug beans, wheat, rye, watercress or radish seed.

PROBLEM:

Rooting (part 1).

NEEDED:

A cutting from a plant, a container of soil, a jelly glass or fruit jar.

DO THIS:

Place the cutting in the soil, water it, and turn the protective glass or jar over it. The cutting should take root and grow.

COMMENT:

This method of starting plants is good for begonias, roses, grapes, tomatoes, azaleas, and some other plants. The jar placed over the cutting protects it from wind or other damage, and helps prevent it from drying out too quickly.

PROBLEM:

Rooting (part 2).

NEEDED:

A plant for rooting, a glass of water, preferably placed in a window.

DO THIS:

Remove the lower leaves from the cutting. Place it in the water and in the light. Add water as necessary, and watch the roots grow! After the roots have developed, the cutting can be potted.

COMMENT:

Swedish ivy, wandering Jew, geraniums, and other plants can be started this way.

PROBLEM:

Moon influence.

NEEDED:

Cuttings for propagation by rooting.

DO THIS:

Follow the old-timers' rule of taking cuttings between a full moon and a new moon. Take others when the moon is in other phases. See whether one group roots more quickly and better than the other.

COMMENT:

The subject of growing crops or flowers "by the moon" is considered important by many. To some, it smacks of astrology, but should perhaps merit more scientific investigation than it gets. Sometimes an old wives' tale turns out to have scientific validity.

PROBLEM:

Plants' breath.

NEEDED:

Seeds, a jar, a vial or small dish of limewater.

DO THIS:

Place seeds in the jar, put some water into the jar so the seeds will germinate, and after a day put the open container of limewater into the jar. It should turn milky, showing that there is carbon dioxide in the jar.

COMMENT:

The rule is that plants give off oxygen as they breathe and take in carbon dioxide. This is true in green plants; they give off much more oxygen than carbon dioxide as they use their chlorophyll to produce food. But it is not true of seedlings. They have no chlorophyll, therefore the foodmaking cannot take place, and they cannot give off oxygen.

This experiment will take a few days.

PROBLEM:

Soil breath.

NEEDED:

Garden soil, a jar, a vial or small dish of limewater (from the drug store).

DO THIS:

Put soil (a few tablespoonsful) into the jar, and set the limewater beside the soil. Cover the mouth of the jar tightly. After a day or two the limewater should turn milky, showing the presence of carbon dioxide in the air in the jar.

HERE'S WHY:

The soil contains countless numbers (millions!) of microscopic plants and animals. They breathe as they live, and their breath is mostly carbon dioxide, which will turn the limewater milky.

The invisible creatures in the soil include bacteria, molds, yeasts and protozoa, Nematodes are sometimes large enough to be seen, and they, too, are numbered in the millions. They are threadlike white worms.

PROBLEM:

Vine without soil.

NEEDED:

A sweet potato, a glass of water.

DO THIS:

Place the potato in the glass, so it rests on the rim with the small end of the potato extending down into the water. Replenish the water when needed. A beautiful vine will grow from the top of the potato.

HERE'S WHY:

The potato, whether in the ground or in a glass, is made up of the food elements needed for the growing plant.

The vine grown from a glass can hang down or can be tied with strings so it extends along a wall or window. It grows better if it can get some sunlight. Select a fresh-looking potato, one that shows the beginnings of buds and roots.

A little fertilizer in the water may improve the growth.

PROBLEM:

The plant waterer.

NEEDED:

A plant, a jar for water, a wire, a cloth.

DO THIS:

Wrap the cloth around the wire so it will be stiff. Put one end into the dirt in the plant pot, and put the other end into the jar of water. See if the cloth will transfer enough water to keep the plant watered.

COMMENT:

Water molecules cling to the tiny fibers of cloth, and this capillary action causes the water to move up through the cloth—against gravity. At the top the water goes down through the cloth by both gravity and capillary action. There should be enough water transfer to keep the plant watered. But try it carefully. Some cloth is treated so that water does not move freely through it.

More cloth wound around the wire will allow more water transfer. Strips from an old sheet were used by the author successfully. Try this out for a week before going away and leaving this watering system to take care of plants. Try three strips of old sheet plaited together. Any wick should reach the bottom of the pot.

PROBLEM:

Seeds and freezing.

NEEDED:

Seeds, small envelopes, a deep freezer.

DO THIS:

Put some bean seeds into two envelopes. Put one envelope into the deep freeze; place the other in a room where it will not be disturbed. Do the same with other seeds, such as peas, radishes, clover.

When planting time comes, plant the seeds, and mark the planting place so you can tell which seeds were frozen and which were not.

COMMENT:

Most dry seeds are not damaged by freezing, even after long periods of time. A science project may be made of this: find out which seeds can stand freezing and which cannot.

PROBLEM:

Air in plants.

NEEDED:

Plants such as celeries, carrots, apples; tree branches; pliers, a pan of water.

DO THIS:

Hold a piece of the plant under water and squeeze it with the pliers. Air bubbles will rise as the air is squeezed out of the plant.

WHICH PROVES:

Air is almost everywhere, even in substances that seem solid. Not all plant tissue contains the same amount of air. Celeries and apples work well in this experiment.

PROBLEM:

A brick garden.

NEEDED:

A brick, a container of water, grass seed.

DO THIS:

Place the brick in the container, and add water until it is half an inch deep. Sprinkle seeds on the top of the brick.

WHAT HAPPENS:

The brick is porous, and water will rise in it, wetting the top so the seeds can sprout. This rising of the water through an apparently solid substance is called "capillary action." The brick is not solid after all.

Some bricks are glazed. They may not work in this experiment, since the glazed surface seals the brick effectively, preventing the water flow up through it. Soaking the brick in water in which a little soluble fertilizer has been added will be helpful in starting sprouting.

PROBLEM:

Roots and erosion.

NEEDED:

A plant growing in a pot; a water hose.

DO THIS:

Select a plant that has been growing in a pot for several weeks. Shake it out. You'll notice that the elaborate network of roots is holding the soil. Try washing the soil out with a water hose. It may be difficult.

COMMENT:

This demonstrates the ability of plants to hold the soil on hillsides. If vegetation is cut away or plowed away, the rain can wash the soil away.

PROBLEM:

"Wonder water."

NEEDED:

Water to boil, a few jars with tight lids, some tomato and lettuce seeds.

DO THIS:

Boil water for five minutes, then put it into airtight jars to prevent air from redissolving into it. (Preheat jars with hot water so they will not break when filled with boiled water.) Let cool.

Soak some of the tomato and lettuce seeds in the cooled, degassed water overnight; soak other seeds in ordinary tap water. Plant the seeds in two separate, marked containers. As they grow, water one container with the boiled water and the other with regular water.

COMMENT:

Several experiments have been performed by biologists who say boiling the water makes "wonder water" out of it. They say plants grown with it do much better than plants grown the regular way. What do you think?

PROBLEM:

Potato eye.

NEEDED:

A potato, two glasses or jars of water.

DO THIS:

Cut two cubes of potato, one with an eye and the other without an eye. Place them in the glasses, and put in enough water to half cover the potato cubes. Let the experiment stand several days. You will see that the potato with the eye will begin to sprout.

WHY:

The potato eye has all the necessary parts to grow and produce a potato vine when planted in soil. Farmers, in planting potatoes, can cut them up so each piece has an eye. That way one potato can make several plants.

PROBLEM:

Why cultivate?

NEEDED:

Two or three square yards of garden soil.

DO THIS:

Divide the ground into two parts. Leave one undisturbed; scratch or "cultivate" the other. See which will grow plants better.

COMMENT:

The experiment may proceed without waiting for plants to grow. After a rain or watering see which square holds moisture better. Water comes up from the ground by capillary action, and mostly stops when it reaches the loose dirt. In uncultivated, or "hard" dirt water rises to the surface where it can evaporate.

PROBLEM:

Seeds.

NEEDED:

Squash or pumpkin seeds, a drinking glass, sand.

DO THIS:

Fill the glass with sand, and push seeds down against the glass so they can be watched. Be sure some seeds are upright, some upside down, and some on their sides. Keep the sand moist and see what happens.

COMMENT:

The plants will grow upright regardless of the position in which the seeds are planted, although growing upright seems to be more trouble for those not planted upright. Just how gravity affects growing plants is not thoroughly understood.

PROBLEM:

Seeds must breathe.

NEEDED:

Four small twist-cap jars, water, radish seed.

DO THIS:

Fill one jar with tap water. Boil some water and pour enough in the second jar to fill it; then pour this water back and forth, into and out of this jar and the third one. (Use pot holders while handling the water and the jars so that you don't scald yourself.) Pour some more boiled water carefully into the fourth jar. Put a seed into each jar, then close the lid tightly on the fourth jar.

WHAT HAPPENS:

The seed in the tap water, which has air dissolved in it, should sprout. Boiling removes most of the air, so the seed in the fourth jar should not sprout. Seed in the second jar may sprout, if the pouring back and forth has succeeded in dissolving some air into the water.

PROBLEM:

Green carnations.

NEEDED:

White carnations, green food coloring, a glass of water.

DO THIS:

Color your carnations for St. Patrick's Day by cutting the stems and placing the flowers in colored water. Cells at the stems will absorb the water, which will be forced from cell to cell up the stems and into the petals. The result will be a colored flower.

If you split a carnation stem, placing half in water of one color and the other in clear water or water of another color, a two-colored flower may be produced.

PROBLEM:

A closed-circuit garden.

NEEDED:

A glass or plastic container, stones, garden soil, seeds, water.

DO THIS:

Place the stones in the container, then the soil, then the plant seeds. Water them once and then after several days, once again.

OBSERVE:

The seeds will grow, sustained only by a small amount of water. Water will evaporate from the soil and plants and condense on the container, so very little of it is lost.

PROBLEM:

Lichens.

NEEDED:

Observation.

DO THIS:

Collect lichens from tree trunks or other wood or stone.

COMMENT:

These lacy incredibles are *two* plants living together. They consist of a fungus and an alga, either of which might do very well living alone—but in the case of lichens, they do better growing together.

Chapter 6

Experiments with Chemistry and Physics

PROBLEM:

A mystery flame.

NEEDED:

Baking soda, vinegar, a jar, a candle on a wire.

DO THIS:

Find a place where there is no draft; a draft might blow the experiment away. Mix vinegar and soda; the mixture will produce carbon dioxide gas that will fill the jar. Lower the lighted candle carefully. The flame should continue to burn on top of the gas briefly after the flame has left the candlestick.

HERE'S WHY:

The hot wick will continue to produce gaseous candle wax for a few seconds, and this gas can continue to burn at the top of the carbon dioxide. The flame cannot burn in the gas, but can burn briefly in the air above the gas.

Experiment a little if this does not work the first time.

PROBLEM:

An enzyme at work.

NEEDED:

A tablespoon of sugar, nine tablespoonfuls of water, some yeast.

DO THIS:

Mix the ingredients in a glass, and place in a warm room. In a few hours the process known as fermentation begins. Bubbles of carbon dioxide rise to the surface, and an odor of alcohol may be smelled.

HERE'S WHY:

The yeast plants produce an enzyme called "zymase," which acts as a catalyst to change the sugar to carbon dioxide and alcohol. Enzymes are necessary to human life. They change the food in our bodies into substances that can nourish us.

In breadbaking the bubbles of gas make the dough rise. The alcohol produced boils away, and the yeast plants are killed. In winemaking the alcohol is utilized.

PROBLEM:

Bernoulli.

NEEDED:

Paper, a ruler, glue.

DO THIS:

Glue the paper to the ruler as shown, and place the ruler on a pencil so the paper end rests against the table. Blow over the ruler and the paper end will rise.

HERE'S WHY:

Bernoulli discovered that air in motion has less pressure than air beside it either at rest or in less rapid motion. Because the air over the paper is flowing faster than the air below it when you blow over the paper, there is less pressure above than below, so the paper end rises.

This is the principle that makes airplanes fly. Air passing over the top of the wings moves more rapidly than the air below, so the greater pressure of the air below keeps the plane up.

PROBLEM:

Why the salt?

NEEDED:

A home ice cream freezer set to go.

DO THIS:

Note that no freezing takes place unless salt is poured in on top of the crushed ice.

HERE'S WHY:

The ice may be only 32 degrees and could not cool the ice cream mixture enough to freeze it. But when salt is added to the ice, its freezing point is lowered, so as it is melted it draws heat from the cream mixture. The temperature of the ice and salt mixture may go as low as 0 degrees F, which is considerably below the melting point of normal ice.

NOTE: This is probably how the original 0 point on the Fahrenheit scale was defined: the lowest temperature obtained with a salt-ice mixture.

PROBLEM:

Elasticity.

NEEDED:

Long strips of scrap glass from a hardware store (scrap may be free), books, gloves, and protective goggles.

DO THIS:

Place two similar strips side by side across books. Press down on the middle of one. It will bend, then return to its original shape. If bent too far it will break.

COMMENT:

Glass is very elastic. That is, it returns to its original shape after being bent, compressed, squeezed, or stretched. Rubber is more stretchable but does not return to its original shape when stretched far and often. If this definition is followed, glass is more elastic than rubber.

Work gloves and goggles will protect in case the glass breaks. Always be careful.

PROBLEM:

Darkening of potatoes.

NEEDED:

Old potatoes to boil, a slice of lemon.

DO THIS:

Boil old potatoes; they will likely turn dark. Try boiling old potatoes, but add a slice of lemon. They should stay white.

HERE'S WHY:

Lemon juice contains citric acid. The acid does the trick.

EXPERIMENT 2:

Fresh fruit should not turn dark if lemon juice is applied to it.

EXPERIMENT 3:

Vinegar, also acid, will help preserve the bright red color of beets.

EXPERIMENT 4:

Slices of fresh fruit dipped in salt water should not turn dark.

It is exposure to oxygen of the air that makes the dark color. These methods help keep oxygen away from the vegetable surfaces.

PROBLEM:

Rocks.

NEEDED:

Several small rocks, preferably from a creek bed, a magnifying glass, cloth and hammer.

DO THIS:

Wrap a small rock in cloth and break it with a hammer. The cloth will keep small pieces from flying. Examine the rock surfaces with the glass.

YOU WILL FIND:

The freshly broken surfaces will be rough and show crystalline shapes. The surfaces that were outside will be more smooth. This is because the motion of rocks and sand over each other in nature tends to wear the surfaces down, making them smoother. The wearing can proceed down until the particles left are sand.

PROBLEM:

Resonating pendulums.

NEEDED:

Four objects of equal weight, string, two chairs.

DO THIS:

Tie a string between the chair backs, then tie two weights to the string with other strings. Make the strings the same length. Then tie two shorter strings and weights as shown.

WHAT HAPPENS:

When one of the longer strings is pulled back and let go, the other longer string takes up the energy and begins to swing. The shorter ones remain almost still. If a short string is allowed to swing, the other shorter weight and string will take up the energy and swing, while the long strings remain still.

A pendulum resonates at one frequency, only, so when one pendulum is set in motion only a pendulum of the same frequency can respond to it.

PROBLEM:

Moments of inertia (part 1).

NEEDED:

Two cardboard disks, one 3 inches and one 4 inches in diameter, straight wooden dowels or round pencils, four yardsticks or meter sticks, books, four drinking glasses alike, rubber bands.

DO THIS:

Mount the disks on the dowels, careful to put them in the centers of the dowels. Mount the yardsticks as shown, making an incline. Let the disks roll down.

WHAT HAPPENS:

The "moment of inertia" is greater in the larger disk, so it starts slower and rolls down more slowly. Moment of inertia is the measure of the resistance of a body to angular or turning acceleration.

PROBLEM:

Maximum moment of inertia (part 2).

NEEDED:

A stick tied to a string.

DO THIS:

Hold the end of the string and rotate it with the hand. Experiment with different ways of rotating the string.

WHAT HAPPENS:

If the string is rotated just right, the stick will rise to a position almost horizontal, almost parallel to the floor. It assumes this position because it tends to rotate about its axis of maximum moment of inertia.

COMMENT:

This is a simple science trick with a technical explanation. "Moment of inertia" may be defined as "the measure of a turning body's resistance to change in the rate of its motion" (Basic Dictionary of Science, E. C. Graham, published by Macmillan).

PROBLEM:

A compressionable wave model.

NEEDED:

Dominoes.

DO THIS:

Place the dominoes in a line, with spaces between them. Push the domino at the end of the line toward the others until they all have come together. Note that the gap closing process (the wave) moves at a much higher velocity or speed than the hand. This is an illustration of a compressionable wave.

The dominoes may be stood upright, but are more likely to stay put if used on their sides.

AN EXAMPLE:

Place five dominoes end to end with space between them of one-fourth the length of a domino. The wave will have moved the full length of all dominoes when the first one has moved its own length, because all spaces will be closed.

PROBLEM:

Particle size.

NEEDED:

Two glasses of water, a lamp, cardboard, sugar, soap.

DO THIS:

Set up the card beside the lamp. Put a spoonful of sugar in one glass; dip a bar of soap into the other. The light should shine through the glass of sugar water.

WHAT HAPPENS:

The sugar water remains clear. Now shine the beam through the other glass. The path of the beam is seen as light is reflected from the soap particles. No reflection is seen from the sugar particles—they are too small.

PROBLEM:

Action and reaction.

NEEDED:

A heavy ball and a rope swing.

DO THIS:

Have a friend stand in front of you. Hold on to the swing rope with one hand, and throw the ball to the friend with the other. Throwing the ball forward gives the swing a push backward.

HERE'S WHY:

It is the principle that makes jets and rockets fly. "For every action there is an equal and opposite reaction." The ball is thrown—the action; the swing is pushed back—the reaction. This is one of Newton's laws.

The person in the swing must keep feet off the ground.

PROBLEM:

Resonance.

NEEDED:

A brick or other heavy object—suspended on a cord.

DO THIS:

Put your mouth near the object and blow on it. The weight should move slightly. It will swing back. Blow again as it begins to move away from the mouth. Repeat blowing, spacing the puff just right, and the weight can be made to swing in a large arc.

COMMENT:

Every object has a natural swing-time or resonant frequency. In this case the puffs of air are timed to the resonant frequency of the swinging pendulum.

Each time you blow on the object in the direction it is moving causes it to move more than on the last swing. The time of the swing depends on the length of the cord and gravity. Pushing a child in a rope swing is another example of resonance.

PROBLEM:

Pull!

NEEDED:

A rope and two people.

DO THIS:

Have someone pull the rope with you. See how strong the pull is. Now tie one end of the rope to a post and pull again. Other factors being equal, the pull with a person on the other end of the rope will equal the pull when one end is tied to the post.

COMMENT:

This illustrates Newton's law of action-reaction. The tension in the rope is the same as the pull force of each person, not the sum of the two pull forces, as someone might think. When the rope is tied to the post, the post pulls on the rope with the same force as the person exerts.

PROBLEM:

Show convection currents.

NEEDED:

A half-pint Mason jar with metal lid, two medicine droppers, a large container for water, a drill, food coloring.

DO THIS:

Cut the tips off of the rubber on the droppers. Drill two holes in the lid and insert the rubber into the holes as shown. Fill the container with cold water. Fill the small jar with very hot water colored deeply. Let the small assembly down into the large container of water.

WHAT HAPPENS:

Hot water is lighter than cold water and will begin to flow up the tube and spread out at the top of the cold water, while cold water flows down the other tube from the large container to the smaller one.

PROBLEM:

Weight.

NEEDED:

A container of water, a piece of wood, a scale.

DO THIS:

Place the container on the scale and fill to overflowing with water. See what the weight is. Then float the wood on the water. Check the weight.

COMMENT:

The wood displaces exactly enough water to weigh the same as the wood. So the weights should be the same.

As a variation, have the container half full of water, then float the wood on it. The weight will increase an amount equal to the weight of the wood, since no water is lost. Look up Archimedes' principle.

PROBLEM:

Tiny explosions.

NEEDED:

Table salt, a pan, heat.

DO THIS:

Sprinkle a little salt in the bottom of the pan, and heat the pan. Soon a crackle and pop will be heard.

WHY:

The salt crystals have water locked up in them. As the water changes to steam, the pressure of the steam bursts the crystals. We hear tiny steam explosions.

PROBLEM:

Sulfur from rubber.

NEEDED:

An aluminum pan, a rubber band or two, a silver spoon (not Mother's best spoon), water, baking soda, vinegar, salt.

DO THIS:

Wrap the band around the bowl of the spoon and leave it overnight. Dark tarnish will form as sulfur from the rubber unites with the silver.

NOW DO THIS:

Clean the tarnish off by putting the silver in an aluminum pan containing water and small amounts of the soda. Boil for a few minutes. Chemical action produces charged ions and this cleans the silver, leaving the aluminum darkened. The aluminum can then be cleaned by boiling vinegar and water in it. There is considerable chemistry here.

PROBLEM:

Inertia: Newton's First Law of Motion.

NEEDED:

A friend on roller skates, a street curb.

DO THIS:

Have the friend skate toward the curb, but squat before reaching it. As the skate is stopped by the curb, the friend tends to continue. He will tumble.

HERE'S WHY:

Newton's First Law: A moving object tends to travel in a straight line, unless acted on by an external force. Here, an obstacle stops the feet, yet the upper part of the body continues to travel.

PROBLEM:

Inertia and Newton's First Law.

NEEDED:

A friend, a soft mat or grass, a beach towel.

DO THIS:

Have the friend squat on the towel on the mat. Jerk the towel quickly; the friend does not come with it, but will tumble over.

HERE'S WHY:

Newton's First Law of Motion: a body at rest will remain at rest unless acted on by an exterior force. In this case, the friction between the friend and the towel is sufficient to move the friend—just enough to make him tumble. He will not move far, forward or backward, from his original position.

PROBLEM:

What stands up when it moves and falls down when it stops?

ANSWER:

A hoop or a wheel.

COMMENT:

This is one of those simple things in science that calls very profound concepts into play. The hoop or wheel becomes a gyroscope when it rolls. Many concepts are at work here: high mathematics, angular velocity, angular momentum, torque and angular acceleration, moment of inertia, and precession.

Explain all this when someone asks what makes the bicycle stand up!

PROBLEM:

A cloth test.

NEEDED:

A lighted candle, cloth samples to test, tweezers.

DO THIS:

Tear or cut very small pieces of the cloth. Hold them with the tweezers in the flame. Cotton and linen burn almost like paper. Wool burns more slowly and smells like burning hair. Nylon may not burn at all, but melts and darkens. Rayon melts after flaming. Pure silk burns slowly and smells like burning hair.

Always be careful of fire; even a small flame can be dangerous. Any experiment with fire should be performed with an adult present.

PROBLEM:

Newton's Third Law.

NEEDED:

A toy locomotive and a section of track; two round pencils.

DO THIS:

Place the track on a smooth board, place the locomotive on it, and place two round pencils under the board. Turn on the electricity; the engine moves forward, but the board and track move backward.

HERE'S WHY:

Newton's third law of motion is: "Every action has an equal but opposite reaction." Here, the board pushes against the engine, but the engine pushes back against the boards.

PROBLEM:

Elastic glass.

NEEDED:

Large plate-glass window.

DO THIS:

Watch the reflection in the window. Press against the glass with a finger, and the reflection will move, showing that the glass has been distorted somewhat by the small pressure.

COMMENT:

Glass is elastic, which means it returns to its original form after being pushed or bent. There are limits, however. If distorted too much, the glass will break.

PROBLEM:

Air and weight.

NEEDED:

Two paper bags, open, balanced on a yardstick.

DO THIS:

Hold a lighted candle under one of the bags. It will rise.

QUESTION:

Did the bag rise because of the draft of air created by the candle flame?

COMMENT:

That is partially true. But do it again, this time holding the bag so it will not rise until the flame is removed. That bag will still rise, showing that warm air in that bag is lighter than air at room temperature in the other bag.

PROBLEM:

Heat and color.

NEEDED:

A tin can, paint, two thermometers, newspaper, a strong light.

DO THIS:

Remove the label from the can, and paint half of it dull black and the other half white. Place the thermometers inside the can, held in place with crumpled newspaper. Shine the light on the can where the black and white parts join.

WHAT HAPPENS:

The black color absorbs more light than the white, changing much of it into heat. The difference in temperature is easily seen.

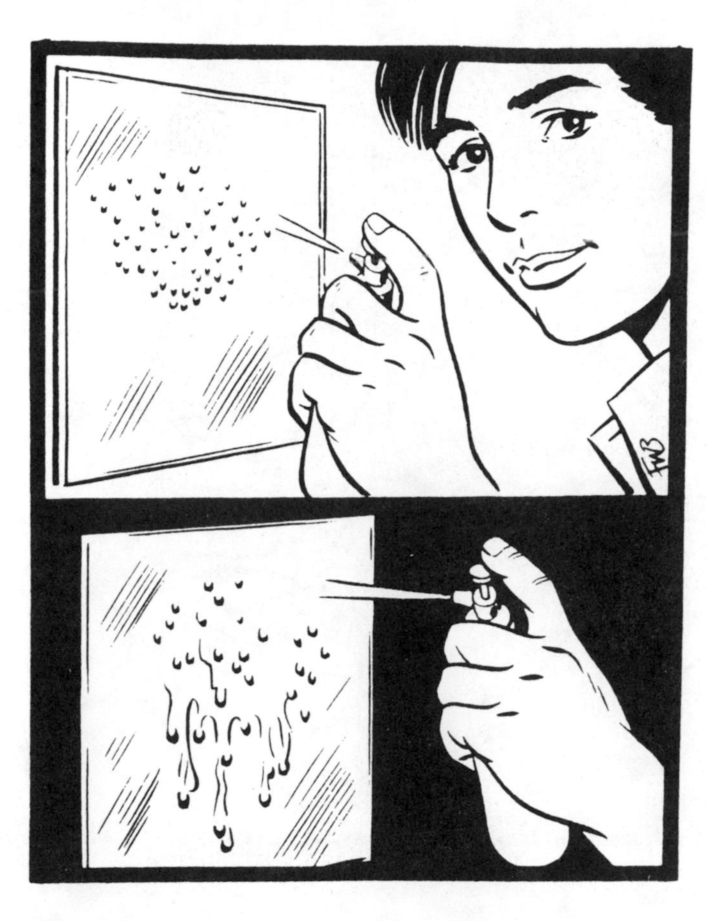

PROBLEM:

Soapy smear.

NEEDED:

Pieces of flat glass or dinner plates, water, detergent, two nonaerosol spray bottles.

DO THIS:

Put water into both bottles. Add a little liquid detergent to one. Spray the water on one glassy surface, the detergent water on another. The plain water tends to form drops, while the detergent water smears and drips and runs off.

HERE'S WHY:

Surface tension tends to make a liquid assume the shape with the smallest surface, the sphere. But the detergent weakens the effect of the surface tension.

PROBLEM:

Convection.

NEEDED:

Two fruit jars, cardboard.

DO THIS:

Heat one of the jars by rinsing it with very hot water. Place it on a table, with a cardboard square on top of it. Place the other jar on top of the card. Lift the card enough for someone to blow smoke into the hot jar, then take the card out, leaving one jar above the other.

WHAT HAPPENS:

Air in the bottom jar, heated by the jar walls, expands so it is lighter than the air in the cold jar. Smoke and air will flow up into the top jar by convection.

A VARIATION:

If the hot jar is on top, nothing happens. The lighter air is already in the upper jar.

PROBLEM:

A siphon.

NEEDED:

A short hose, two jars, some water.

DO THIS:

Fill a jar with water. Fill the hose with water. Insert the hose into the jars, positioning one higher than the other.

WHAT HAPPENS:

Water will flow through the hose from the higher jar to the lower jar. This is the principle of the siphon.

To start the siphon, fill the hose with water, hold the ends as they are placed in the jars. The back-and-forth flow can be kept up indefinitely, but if a hose end slips out of the water and sucks air, the action stops. Atmospheric pressure on the top of the water pushes the flow from upper to lower containers.

PROBLEM:

An emulsion you can eat.

NEEDED:

Two eggs, one cup of olive oil or other good food oil, ¼ teaspoon of salt, two teaspoons vinegar, a bowl, and an egg beater.

DO THIS:

Separate egg yolks from egg whites. (You will not need egg whites.) Beat the egg yolks and salt together, add half the vinegar and a third of the oil, a few drops at a time, beating constantly. Add the rest of the oil slowly, continue beating until the mixture is thick. Beat in the rest of the vinegar and put the mixture, which is an emulsion, into the refrigerator.

WHAT HAS HAPPENED:

Oil and vinegar ordinarily would not mix, but the egg yolk acts as an "emulsifying agent," joining them together into a mixture which most people consider delicious to eat. We call it "mayonnaise."

NOTE:

The eggs and other ingredients should be at room temperature before the experiment is started. Use good materials to get a good taste.

PROBLEM:

Moire patterns.

NEEDED:

Two combs.

DO THIS:

Hold the combs together as shown. You will see beautiful "moire patterns."

HERE'S WHY:

The patterns are formed either by the light that passes through the combs or by dark areas where light does not pass through. Brilliance and beauty are enhanced because of diffraction of light at the small comb openings.

Try two pieces of screen wire and other objects. The name "moire" means "watered." The effect is related to some effects in water and cloth. Pronounce the word mow-ray.

PROBLEM:

Radiation and convection.

NEEDED:

A sauce pan of hot water and someone to hold it.

DO THIS:

Hold one hand above the pan and one hand below it at equal distances from the water. Heat will be felt below the pan, but more heat will be felt above it.

HERE'S WHY:

The heat that is felt below the pan comes down by radiation. This radiation is similar to visible light and is called infrared. The heat that is felt above, comes partially from radiation, but mostly by convection.

This means that the air and warmed vapor rises from the hot water against the hand. The vapor rises when heated because it expands and becomes lighter than air at normal temperature.

PROBLEM:

Solubility.

NEEDED:

Salt, sugar, water, a pan, heat, a spoon, a glass.

DO THIS:

Fill the glass half full of water, and add sugar until the water will dissolve no more. Warm the solution, and you will see that more sugar will dissolve. When you try the same procedure using salt instead of sugar, notice that practically no more will dissolve in warm water than in cold.

HERE'S WHY:

Different compounds dissolve at different rates as the temperature increases. Sugar is a compound that will dissolve more rapidly at higher temperatures than salt.

PROBLEM:

Light reflection and refraction.

NEEDED:

A fish tank, a flashlight.

DO THIS:

Put water in the tank. Add a few drops of milk. Stir. Hold the flashlight so the beam enters the water at various angles. Put smoke into the air above the tank.

WHAT HAPPENS:

The light bends or refracts as it moves from air to and through water. The strange angles of the light rays are quite interesting. The smoke particles and milk particles let the light paths show. If a sheet of glass is used to cover the tank even more reflections and refractions are seen.

Index